COMPUTER TESTING SUPPLEMENT
FOR
COMMERCIAL PILOT

1998

U.S. DEPARTMENT OF TRANSPORTATION
FEDERAL AVIATION ADMINISTRATION

Flight Standards Service

PREFACE

This computer testing supplement is designed by the Flight Standards Service of the Federal Aviation Administration (FAA) for use by computer testing designees (CTD's) and testing centers in the administration of airman knowledge tests in the following knowledge areas:

Commercial Pilot—Airplane (CAX)
Commercial Pilot—Glider (CGX)
Commercial Pilot—Lighter-Than-Air–Airship (CLA)
Commercial Pilot—Rotorcraft/Gyroplane (CRG)
Commercial Pilot—Rotorcraft/Helicopter (CRH)
Commercial Pilot—Free Balloon Gas (CBG)
Commercial Pilot—Free Balloon–Hot Air (CBH)
Military Competence—Airplane (MCA)
Military Competence—Rotorcraft/Helicopter (MCH)

Comments regarding this supplement should be sent to:

U.S. Department of Transportation
Federal Aviation Administration
Flight Standards Service
Airman Testing Standards Branch, AFS-630
P.O. Box 25082
Oklahoma City, OK 73125

CONTENTS

CONTENTS—Continued

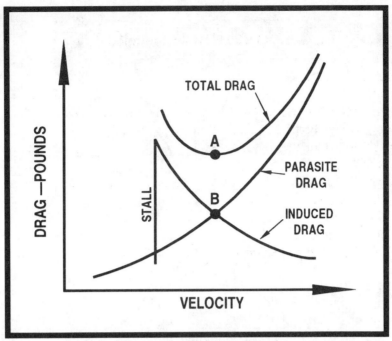

FIGURE 1.—Drag vs. Speed.

GROSS WEIGHT 2750 LBS		ANGLE OF BANK			
		LEVEL	30°	45°	60°
POWER		GEAR AND FLAPS UP			
ON	MPH	62	67	74	88
	KTS	54	58	64	76
OFF	MPH	75	81	89	106
	KTS	65	70	77	92
		GEAR AND FLAPS DOWN			
ON	MPH	54	58	64	76
	KTS	47	50	56	66
OFF	MPH	66	71	78	93
	KTS	57	62	68	81

FIGURE 2.—Stall Speeds.

1

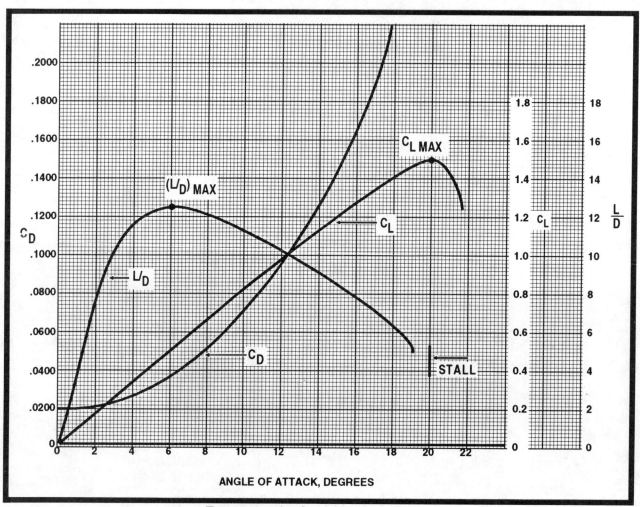

FIGURE 3.—Angle of Attack, Degrees.

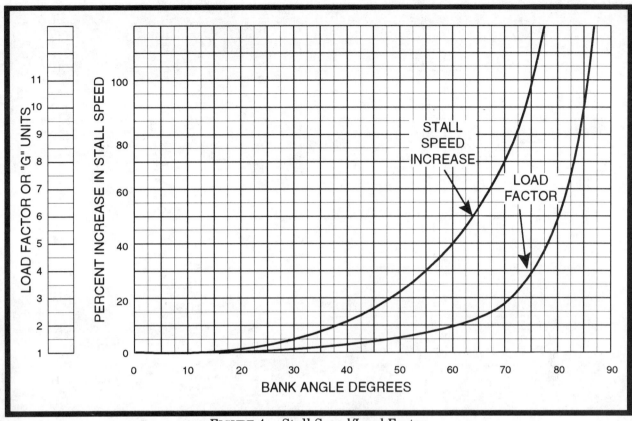

FIGURE 4.—Stall Speed/Load Factor.

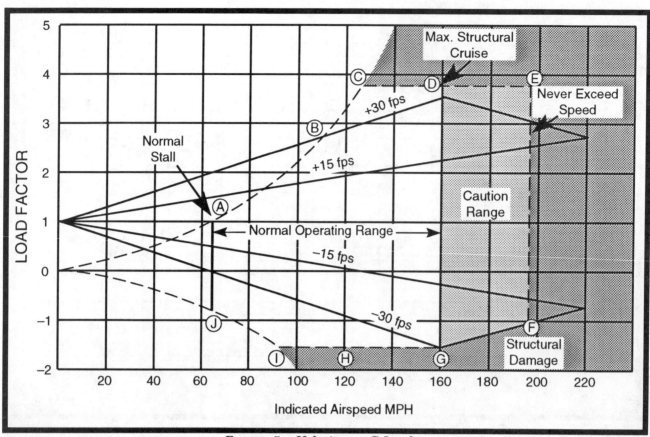

FIGURE 5.—Velocity vs. G-Loads.

4

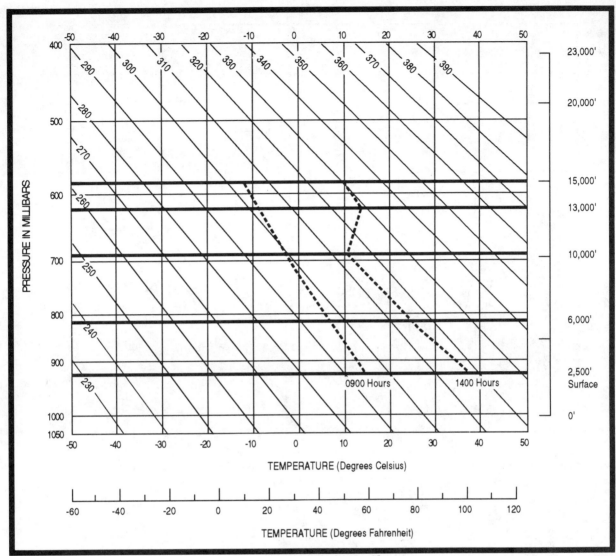

FIGURE 6.—Adiabatic Chart.

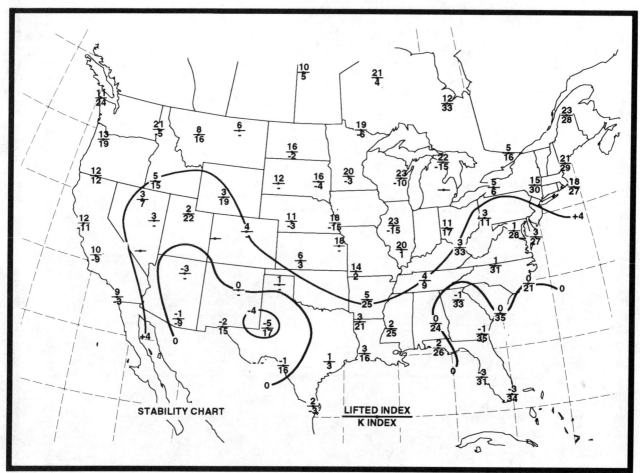

FIGURE 7.—Stability Chart.

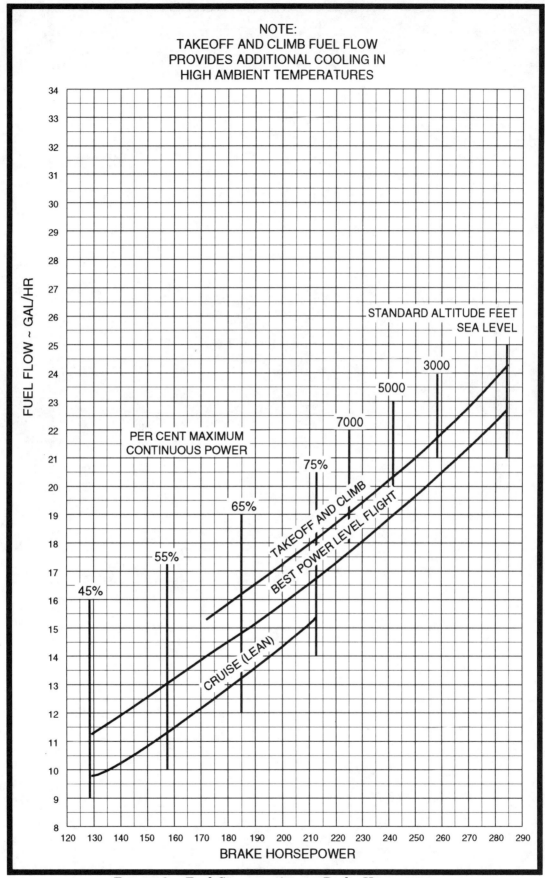

FIGURE 8.—Fuel Consumption vs. Brake Horsepower.

NORMAL CLIMB – 100 KIAS

CONDITIONS:
Flaps Up
Gear Up
2550 RPM
25 Inches MP or Full Throttle
Cowl Flaps Open
Standard Temperature

MIXTURE SETTING	
PRESS ALT	PPH
S.L. to 4000	108
8000	96
12,000	84

NOTES:
1. Add 12 pounds of fuel for engine start, taxi and takeoff allowance.
2. Increase time, fuel and distance by 10% for each 10 °C above standard temperature.
3. Distances shown are based on zero wind.

WEIGHT LBS	PRESS ALT FT	RATE OF CLIMB FPM	FROM SEA LEVEL		
			TIME MIN	FUEL USED POUNDS	DISTANCE NM
3800	S.L.	580	0	0	0
	2000	580	3	6	6
	4000	570	7	12	12
	6000	470	11	19	19
	8000	365	16	27	28
	10,000	265	22	37	40
	12,000	165	32	51	59
3500	S.L.	685	0	0	0
	2000	685	3	5	5
	4000	675	6	11	10
	6000	565	9	16	16
	8000	455	13	23	23
	10,000	350	18	31	33
	12,000	240	25	41	46
3200	S.L.	800	0	0	0
	2000	800	2	4	4
	4000	795	5	9	8
	6000	675	8	14	13
	8000	560	11	19	19
	10,000	445	15	25	27
	12,000	325	20	33	37

FIGURE 9.—Fuel, Time, and Distance to Climb.

MAXIMUM RATE OF CLIMB

CONDITIONS:
Flaps Up
Gear Up
2700 RPM
Full Throttle
Mixture Set at Placard Fuel Flow
Cowl Flaps Open
Standard Temperature

MIXTURE SETTING	
PRESS ALT	PPH
S.L.	138
4000	126
8000	114
12,000	102

NOTES:
1. Add 12 pounds of fuel for engine start, taxi and takeoff allowance.
2. Increase time, fuel and distance by 10% for each 10 °C above standard temperature.
3. Distances shown are based on zero wind.

WEIGHT LBS	PRESS ALT FT	CLIMB SPEED KIAS	RATE OF CLIMB FPM	FROM SEA LEVEL		
				TIME MIN	FUEL USED POUNDS	DISTANCE NM
3800	S.L.	97	860	0	0	0
	2000	95	760	2	6	4
	4000	94	660	5	12	9
	6000	93	565	9	18	14
	8000	91	465	13	26	21
	10,000	90	365	18	35	29
	12,000	89	265	24	47	41
3500	S.L.	95	990	0	0	0
	2000	94	885	2	5	3
	4000	93	780	5	10	7
	6000	91	675	7	16	12
	8000	90	570	11	22	17
	10,000	89	465	15	29	24
	12,000	87	360	20	38	32
3200	S.L.	94	1135	0	0	0
	2000	92	1020	2	4	3
	4000	91	910	4	9	6
	6000	90	800	6	14	10
	8000	88	685	9	19	14
	10,000	87	575	12	25	20
	12,000	86	465	16	32	26

FIGURE 10.—Fuel, Time, and Distance to Climb.

| | | | | | Gross Weight- 2300 Lbs. Standard Conditions Zero Wind Lean Mixture | | | |

NOTE: Maximum cruise is normally limited to 75% power.

ALT.	RPM	% BHP	TAS MPH	GAL/ HOUR	38 GAL (NO RESERVE)		48 GAL (NO RESERVE)	
					ENDR. HOURS	RANGE MILES	ENDR. HOURS	RANGE MILES
2500	2700	86	134	9.7	3.9	525	4.9	660
	2600	79	129	8.6	4.4	570	5.6	720
	2500	72	123	7.8	4.9	600	6.2	760
	2400	65	117	7.2	5.3	620	6.7	780
	2300	58	111	6.7	5.7	630	7.2	795
	2200	52	103	6.3	6.1	625	7.7	790
5000	2700	82	134	9.0	4.2	565	5.3	710
	2600	75	128	8.1	4.7	600	5.9	760
	2500	68	122	7.4	5.1	625	6.4	790
	2400	61	116	6.9	5.5	635	6.9	805
	2300	55	108	6.5	5.9	635	7.4	805
	2200	49	100	6.0	6.3	630	7.9	795
7500	2700	78	133	8.4	4.5	600	5.7	755
	2600	71	127	7.7	4.9	625	6.2	790
	2500	64	121	7.1	5.3	645	6.7	810
	2400	58	113	6.7	5.7	645	7.2	820
	2300	52	105	6.2	6.1	640	7.7	810
10,000	2650	70	129	7.6	5.0	640	6.3	810
	2600	67	125	7.3	5.2	650	6.5	820
	2500	61	118	6.9	5.5	655	7.0	830
	2400	55	110	6.4	5.9	650	7.5	825
	2300	49	100	6.0	6.3	635	8.0	800

FIGURE 11.—Cruise and Range Performance.

10

PRESSURE ALTITUDE 18,000 FEET

CONDITIONS:
4000 Pounds
Recommended Lean Mixture
Cowl Flaps Closed

NOTE
For best fuel economy at 70% power or less, operate at 6 PPH leaner than shown in this chart or at peak EGT.

RPM	MP	20 °C BELOW STANDARD TEMP -41 °C			STANDARD TEMPERATURE -21 °C			20 °C ABOVE STANDARD TEMP -1 °C		
		% BHP	KTAS	PPH	% BHP	KTAS	PPH	% BHP	KTAS	PPH
2500	30	---	---	---	81	188	106	76	185	100
	28	80	184	105	76	182	99	71	178	93
	26	75	178	99	71	176	93	67	172	88
	24	70	171	91	66	168	86	62	164	81
	22	63	162	84	60	159	79	56	155	75
2400	30	81	185	107	77	183	101	72	180	94
	28	76	179	100	72	177	94	67	173	88
	26	71	172	93	67	170	88	63	166	83
	24	66	165	87	62	163	82	58	159	77
	22	61	158	80	57	155	76	54	150	72
2300	30	79	182	103	74	180	97	70	176	91
	28	74	176	97	70	174	91	65	170	86
	26	69	170	91	65	167	86	61	163	81
	24	64	162	84	60	159	79	56	155	75
	22	58	154	77	55	150	73	51	145	65
2200	26	66	166	87	62	163	82	58	159	77
	24	61	158	80	57	154	76	54	150	72
	22	55	148	73	51	144	69	48	138	66
	20	49	136	66	46	131	63	43	124	59

FIGURE 12.—Cruise Performance.

11

MAXIMUM RATE OF CLIMB

CONDITIONS:
Flaps Up
Gear Up
2600 RPM
Cowl Flaps Open
Standard Temperature

PRESS ALT	MP	PPH
S.L. TO 17,000	35	162
18,000	34	156
20,000	32	144
22,000	30	132
24,000	28	120

NOTES:
1. Add 16 pounds of fuel for engine start, taxi and takeoff allowance.
2. Increase time, fuel and distance by 10% for each 10 °C above standard temperature.
3. Distances shown are based on zero wind.

WEIGHT LBS	PRESS ALT FT	CLIMB SPEED KIAS	RATE OF CLIMB FPM	FROM SEA LEVEL		
				TIME MIN	FUEL USED POUNDS	DISTANCE NM
4000	S.L.	100	930	0	0	0
	4000	100	890	4	12	7
	8000	100	845	9	24	16
	12,000	100	790	14	38	25
	16,000	100	720	19	52	36
	20,000	99	515	26	69	50
	24,000	97	270	37	92	74
3700	S.L.	99	1060	0	0	0
	4000	99	1020	4	10	6
	8000	99	975	8	21	13
	12,000	99	915	12	33	21
	16,000	99	845	17	45	30
	20,000	97	630	22	59	42
	24,000	95	370	30	77	60
3400	S.L.	97	1205	0	0	0
	4000	97	1165	3	9	5
	8000	97	1120	7	19	12
	12,000	97	1060	11	29	18
	16,000	97	985	15	39	26
	20,000	96	760	19	51	36
	24,000	94	485	26	65	50

FIGURE 13.—Fuel, Time, and Distance to Climb.

NORMAL CLIMB – 110 KIAS

CONDITIONS:
Flaps Up
Gear Up
2500 RPM
30 Inches Hg
120 PPH Fuel Flow
Cowl Flaps Open
Standard Temperature

NOTES:
1. Add 16 pounds of fuel for engine start, taxi and takeoff allowance.
2. Increase time, fuel and distance by 10% for each 7 °C above standard temperature.
3. Distances shown are based on zero wind.

WEIGHT LBS	PRESS ALT FT	RATE OF CLIMB FPM	FROM SEA LEVEL		
			TIME MIN	FUEL USED POUNDS	DISTANCE NM
4000	S.L.	605	0	0	0
	4000	570	7	14	13
	8000	530	14	28	27
	12,000	485	22	44	43
	16,000	430	31	62	63
	20,000	365	41	82	87
	S.L.	700	0	0	0
3700	4000	665	6	12	11
	8000	625	12	24	23
	12,000	580	19	37	37
	16,000	525	26	52	53
	20,000	460	34	68	72
	S.L.	810	0	0	0
	4000	775	5	10	9
3400	8000	735	10	21	20
	12,000	690	16	32	31
	16,000	635	22	44	45
	20,000	565	29	57	61

FIGURE 14.—Fuel, Time, and Distance to Climb.

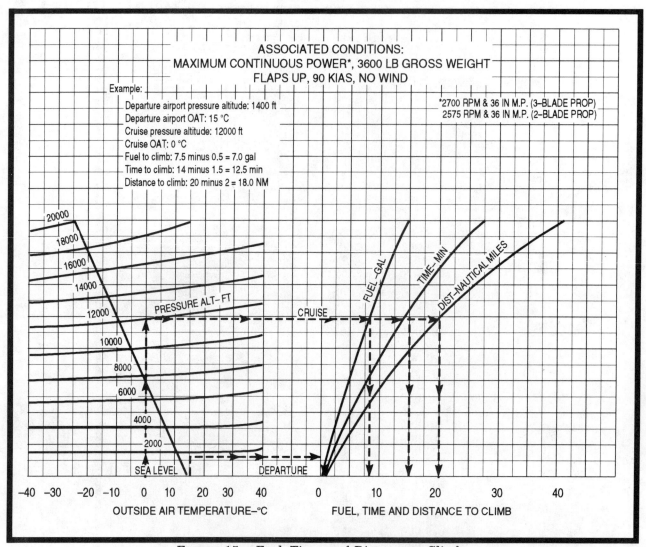

FIGURE 15.—Fuel, Time, and Distance to Climb.

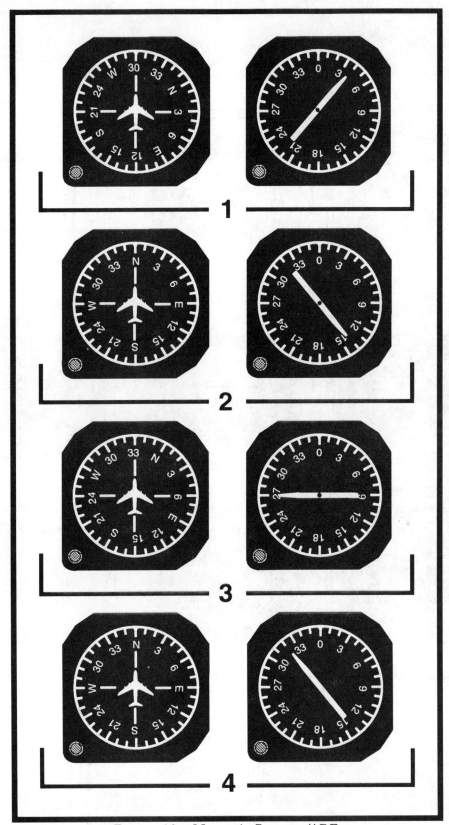

FIGURE 16.—Magnetic Compass/ADF.

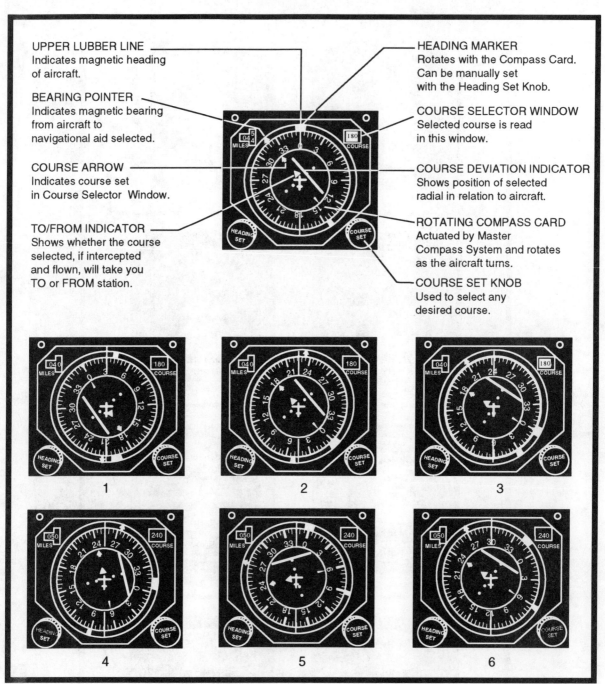

UPPER LUBBER LINE
Indicates magnetic heading
of aircraft.

BEARING POINTER
Indicates magnetic bearing
from aircraft to
navigational aid selected.

COURSE ARROW
Indicates course set
in Course Selector Window.

TO/FROM INDICATOR
Shows whether the course
selected, if intercepted
and flown, will take you
TO or FROM station.

HEADING MARKER
Rotates with the Compass Card.
Can be manually set
with the Heading Set Knob.

COURSE SELECTOR WINDOW
Selected course is read
in this window.

COURSE DEVIATION INDICATOR
Shows position of selected
radial in relation to aircraft.

ROTATING COMPASS CARD
Actuated by Master
Compass System and rotates
as the aircraft turns.

COURSE SET KNOB
Used to select any
desired course.

FIGURE 17.—Horizontal Situation Indicator (HSI).

16

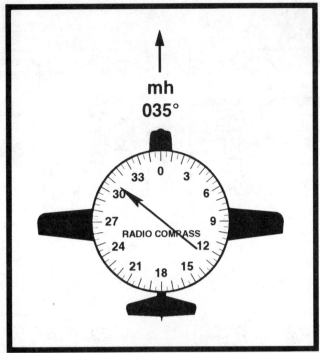

FIGURE 18.—Magnetic Heading/Radio Compass.

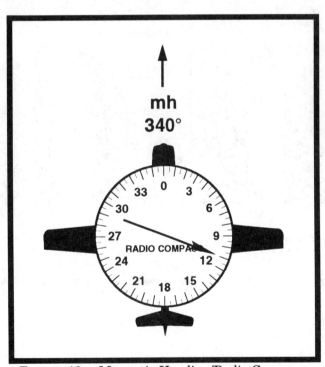

FIGURE 19.—Magnetic Heading/Radio Compass.

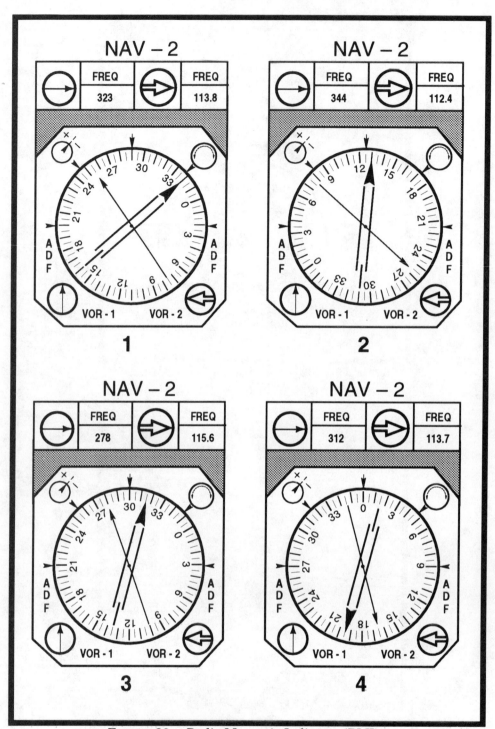

FIGURE 20.—Radio Magnetic Indicator (RMI).

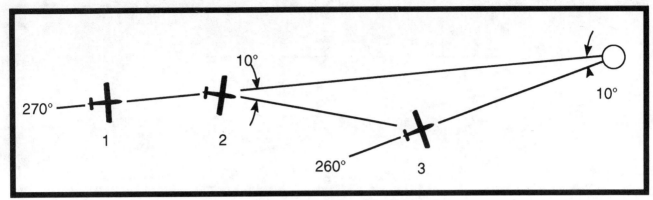

FIGURE 21.—Isosceles Triangle.

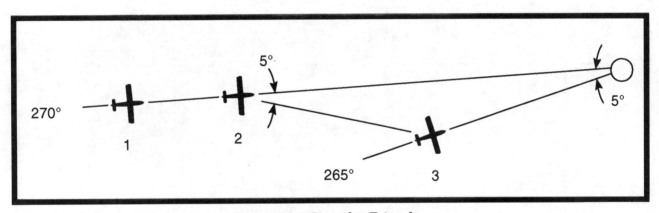

FIGURE 22.—Isosceles Triangle.

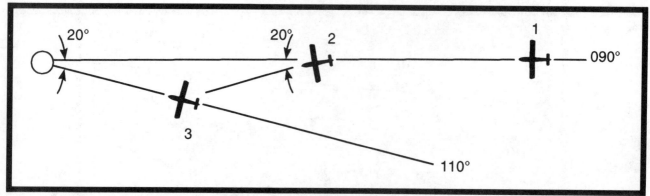

FIGURE 23.—Isosceles Triangle.

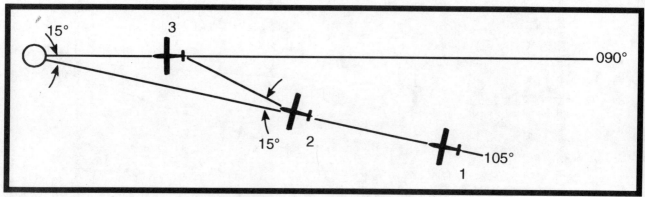

FIGURE 24.—Isosceles Triangle.

20

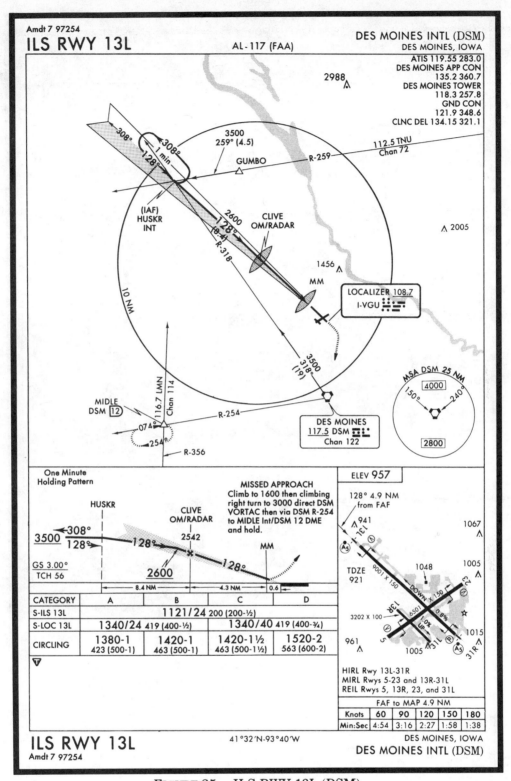

FIGURE 25.—ILS RWY 13L (DSM).

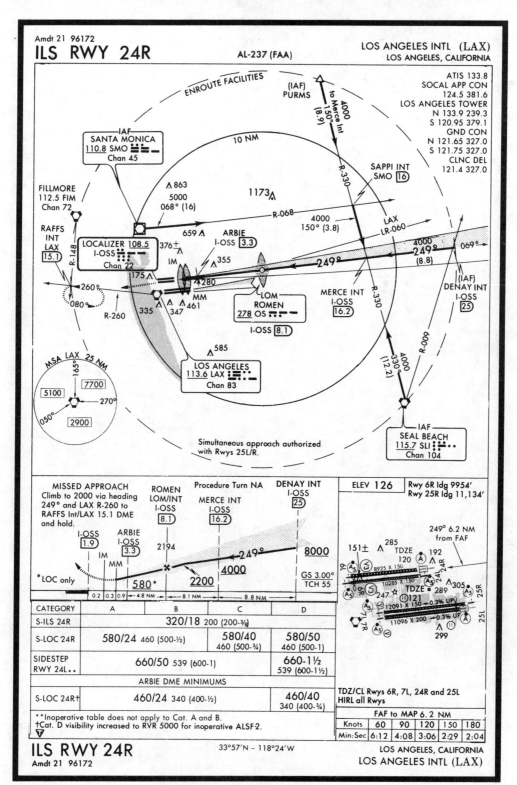

FIGURE 26.—ILS RWY 24R (LAX).

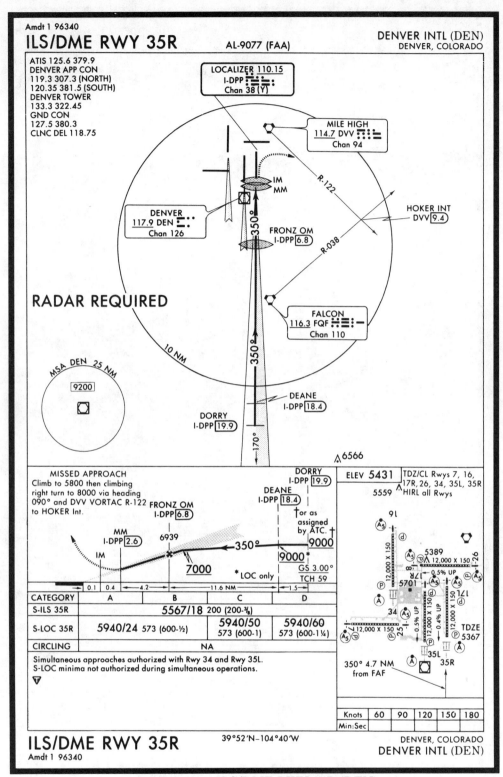

FIGURE 27.—ILS/DME RWY 35R (DEN).

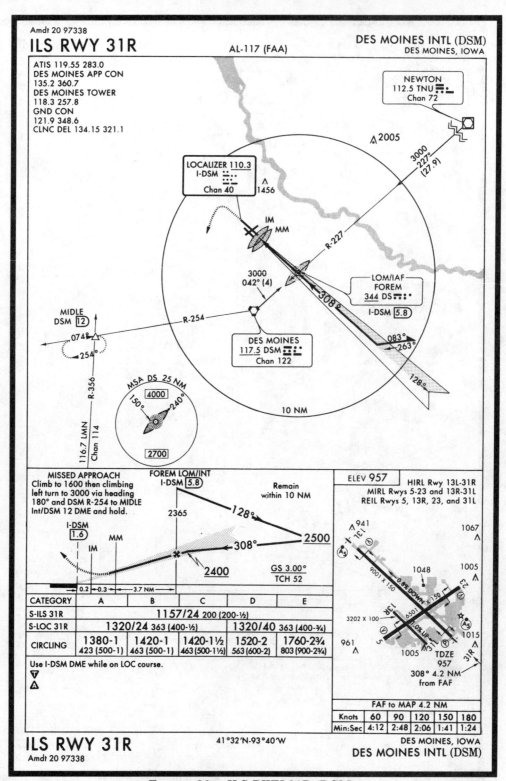

FIGURE 28.—ILS RWY 31R (DSM).

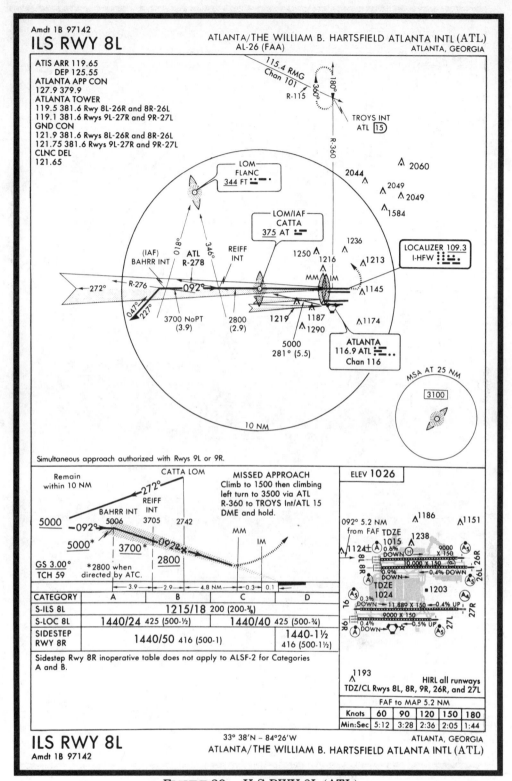

FIGURE 29.—ILS RWY 8L (ATL).

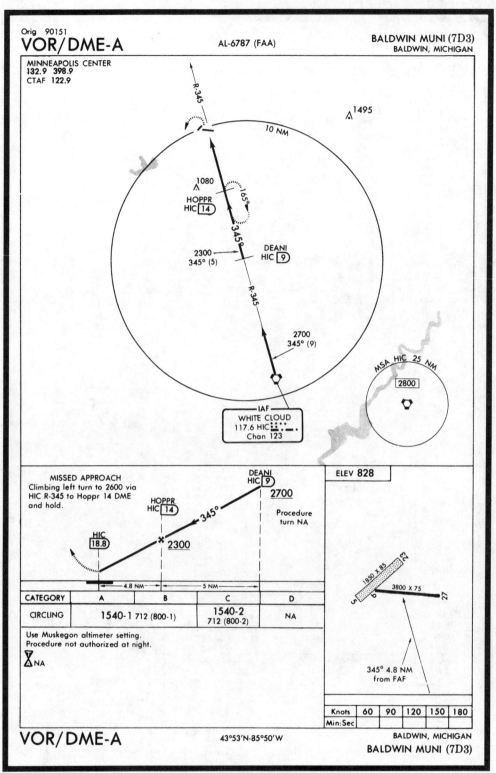

Orig 90151
VOR/DME-A

AL-6787 (FAA)

BALDWIN MUNI (7D3)
BALDWIN, MICHIGAN

MINNEAPOLIS CENTER
132.9 398.9
CTAF 122.9

R-345

△1495

10 NM

△1080

HOPPR
HIC 14

165°

345°

2300
345° (5)

DEANI
HIC 9

R-345

2700
345° (9)

MSA HIC 25 NM

2800

IAF
WHITE CLOUD
117.6 HIC ⋯
Chan 123

MISSED APPROACH
Climbing left turn to 2600 via
HIC R-345 to Hoppr 14 DME
and hold.

DEANI
HIC 9
2700

ELEV 828

HOPPR
HIC 14

345°

Procedure
turn NA

HIC
18.8

2300

4.8 NM

5 NM

1850 X 85

3800 X 75

CATEGORY	A	B	C	D
CIRCLING	1540-1 712 (800-1)		1540-2 712 (800-2)	NA

Use Muskegon altimeter setting.
Procedure not authorized at night.

X NA

345° 4.8 NM
from FAF

Knots	60	90	120	150	180
Min:Sec					

VOR/DME-A

43°53'N-85°50'W

BALDWIN, MICHIGAN
BALDWIN MUNI (7D3)

FIGURE 30.—VOR/DME-A (7D3).

26

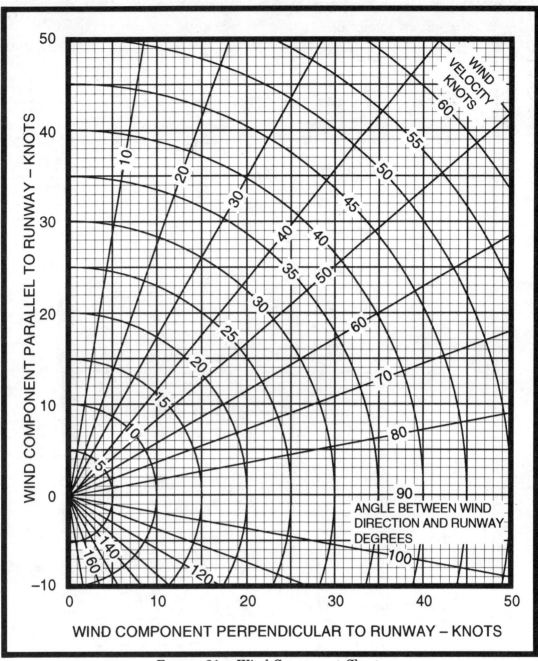

FIGURE 31.—Wind Component Chart.

ASSOCIATED CONDITIONS:

POWER	TAKEOFF POWER SET BEFORE BRAKE RELEASE
FLAPS	20*
RUNWAY	PAVED, LEVEL, DRY SURFACE
TAKEOFF SPEED	IAS AS TABULATED

NOTE: GROUND ROLL IS APPROX. 73% OF TOTAL TAKEOFF DISTANCE OVER A 50 FT OBSTACLE

EXAMPLE:

OAT	75 °F
PRESSURE ALTITUDE	4000 FT
TAKEOFF WEIGHT	3100 LB
HEADWIND	20 KNOTS

TOTAL TAKEOFF DISTANCE OVER A 50 FT OBSTACLE	1350 FT
GROUND ROLL (73% OF 1350)	986 FT
IAS TAKEOFF SPEED	
LIFT-OFF	74 MPH
AT 50 FT	74 MPH

| WEIGHT POUNDS | IAS TAKEOFF SPEED (ASSUMES ZERO INSTR. ERROR) | | | |
| | LIFT-OFF | | 50 FEET | |
	MPH	KNOTS	MPH	KNOTS
3400	77	67	77	67
3200	75	65	75	65
3000	72	63	72	63
2800	69	60	69	60
2600	66	57	66	57
2400	63	55	63	55

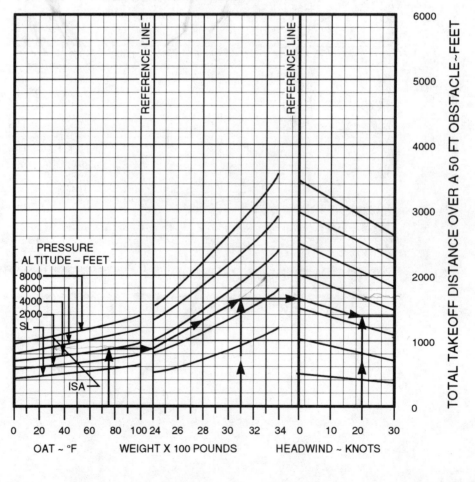

FIGURE 32.—Obstacle Take-off Chart.

CONDITIONS:
Flaps Up
Gear Up
2600 RPM
Cowl Flaps Open

PRESS ALT	MP	PPH
S.L. TO 17,000	35	162
18,000	34	156
20,000	32	144
22,000	30	132
24,000	28	120

WEIGHT LBS	PRESS ALT FT	CLIMB SPEED KIAS	RATE OF CLIMB – FPM			
			-20 °C	0 °C	20 °C	40 °C
4000	S.L.	100	1170	1035	895	755
	4000	100	1080	940	800	655
	8000	100	980	840	695	555
	12,000	100	870	730	590	---
	16,000	100	740	605	470	---
	20,000	99	485	355	---	---
	24,000	97	190	70	---	---
3700	S.L.	99	1310	1165	1020	875
	4000	99	1215	1070	925	775
	8000	99	1115	965	815	670
	12,000	99	1000	855	710	---
	16,000	99	865	730	590	---
	20,000	97	600	470	---	---
	24,000	95	295	170	---	---
3400	S.L.	97	1465	1320	1165	1015
	4000	97	1370	1220	1065	910
	8000	97	1265	1110	955	795
	12,000	97	1150	995	845	---
	16,000	97	1010	865	725	---
	20,000	96	730	595	---	---
	24,000	94	405	275	---	---

FIGURE 33.—Maximum Rate-of-Climb Chart.

29

PRESSURE ALTITUDE 6,000 FEET

CONDITIONS:
Recommended Lean Mixture
3800 Pounds
Cowl Flaps Closed

RPM	MP	20 °C BELOW STANDARD TEMP -17 °C			STANDARD TEMPERATURE 3 °C			20 °C ABOVE STANDARD TEMP 23 °C		
		% BHP	KTAS	PPH	% BHP	KTAS	PPH	% BHP	KTAS	PPH
2550	24	---	---	---	78	173	97	75	174	94
	23	76	167	96	74	169	92	71	171	89
	22	72	164	90	69	166	87	67	167	84
	21	68	160	85	65	162	82	63	163	80
2500	24	78	169	98	75	171	95	73	172	91
	23	74	166	93	71	167	90	69	169	87
	22	70	162	88	67	164	85	65	165	82
	21	66	158	83	63	160	80	61	160	77
2400	24	73	165	91	70	166	88	68	167	85
	23	69	161	87	67	163	84	64	164	81
	22	65	158	82	63	159	79	61	160	77
	21	61	154	77	59	155	75	57	155	73
2300	24	68	161	86	66	162	83	64	163	80
	23	65	158	82	62	159	79	60	159	76
	22	61	154	77	59	155	75	57	155	72
	21	57	150	73	55	150	71	53	150	68
2200	24	63	156	80	61	157	77	59	158	75
	23	60	152	76	58	153	73	56	154	71
	22	57	149	72	54	149	70	53	149	67
	21	53	144	68	51	144	66	49	143	64
	20	50	139	64	48	138	62	46	137	60
	19	46	133	60	44	132	58	43	131	57

FIGURE 34.—Cruise Performance Chart.

ASSOCIATED CONDITIONS:

POWER   AS REQUIRED TO
                  MAINTAIN 800 FT/MIN
                  DESCENT ON APPROACH

FLAPS   DOWN

RUNWAY   PAVED, LEVEL,
                  DRY SURFACE

APPROACH
SPEED   IAS A TABULATED

NOTE: GROUND ROLL IS APPROX. 53%
             OF TOTAL LANDING DISTANCE
             OVER A 50 FT OBSTACLE.

EXAMPLE:

OAT	75 °F
PRESSURE ALTITUDE	4000 FT
LANDING WEIGHT	3200 LB
HEADWIND	10 KNOTS
TOTAL LANDING DISTANCE	
OVER A 50 FT OBSTACLE	1475 FT
GROUND ROLL (53% OF 1475)	782 FT
IAS APPROACH SPEED	87 MPH IAS

WEIGHT POUNDS	IAS APPROACH SPEED (ASSUMES ZERO INSTR. ERROR)	
	MPH	KNOTS
3400	90	78
3200	87	76
3000	84	73
2800	81	70
2600	78	68
2400	75	65

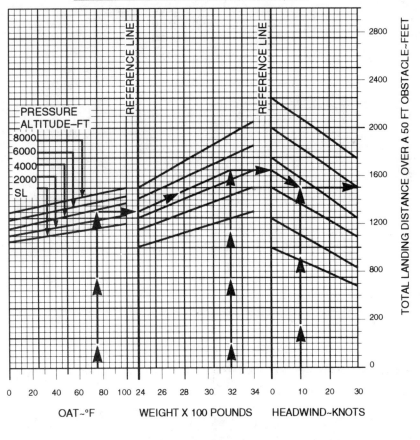

FIGURE 35.—Normal Landing Chart.

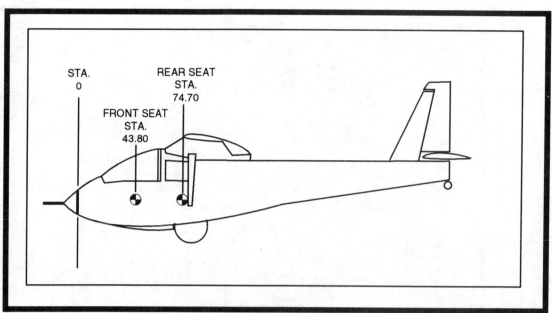

FIGURE 36.—Stations Diagram.

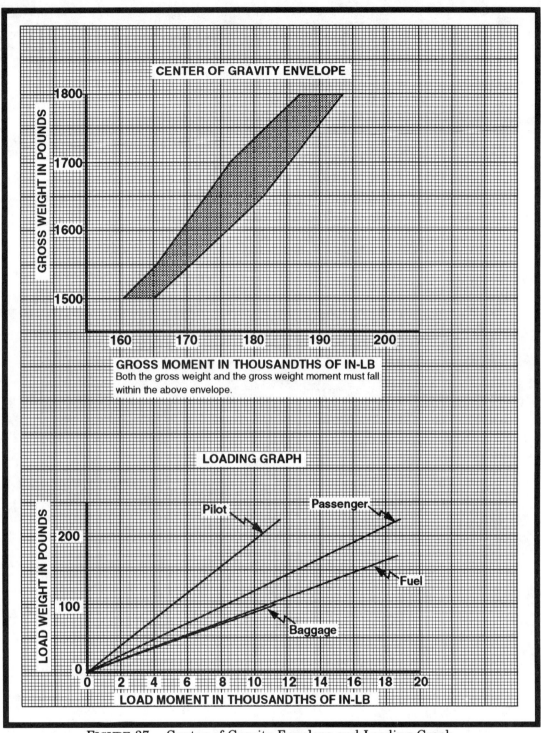

FIGURE 37.—Center-of-Gravity Envelope and Loading Graph.

33

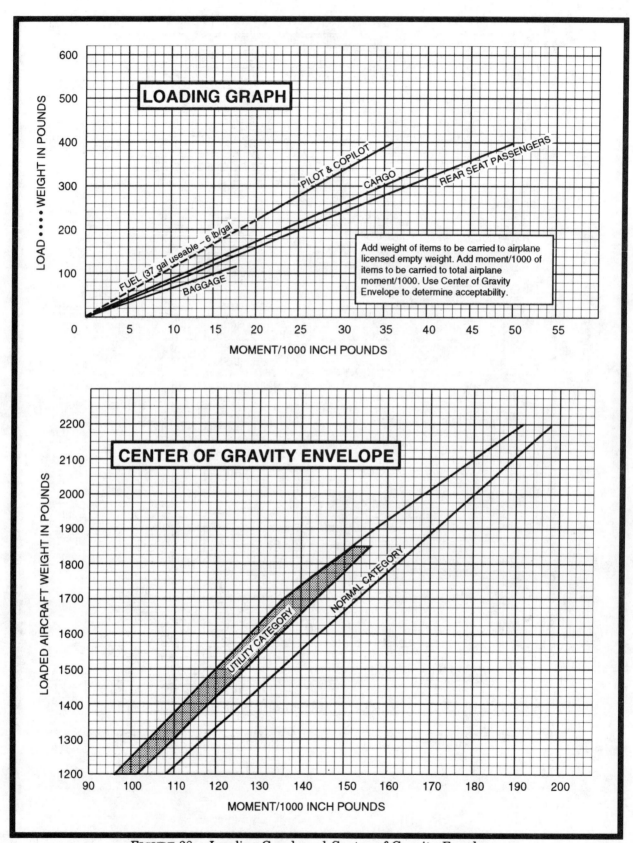

Add weight of items to be carried to airplane licensed empty weight. Add moment/1000 of items to be carried to total airplane moment/1000. Use Center of Gravity Envelope to determine acceptability.

FIGURE 38.—Loading Graph and Center-of-Gravity Envelope.

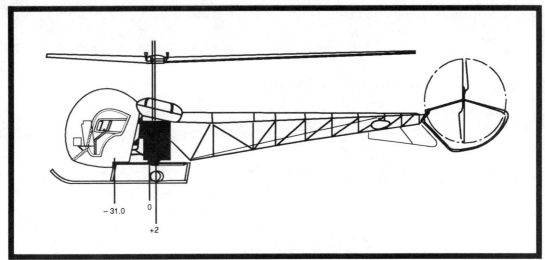

FIGURE 39.—Stations Diagram.

The following CG locations may be used when determining the helicopter CG position.

Item	Long CG	Lat CG
Pilot & Baggage under R seat	79.0	+10.7
Passenger & Baggage under L seat	79.0	−9.3
Main Fuel	108.6	−11.0
Aux Fuel (optional)	103.8	+11.2

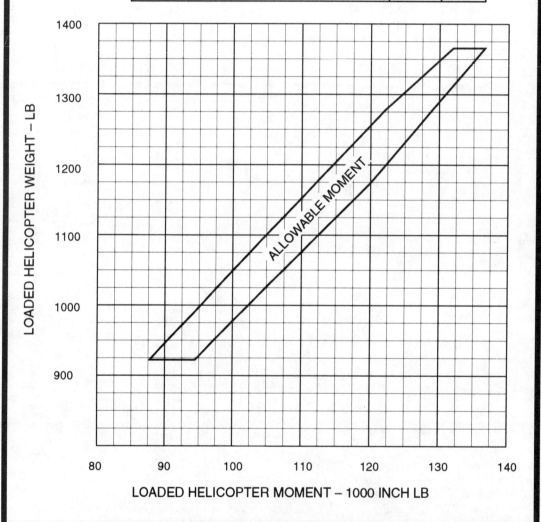

FIGURE 40.—Weight and Balance Chart.

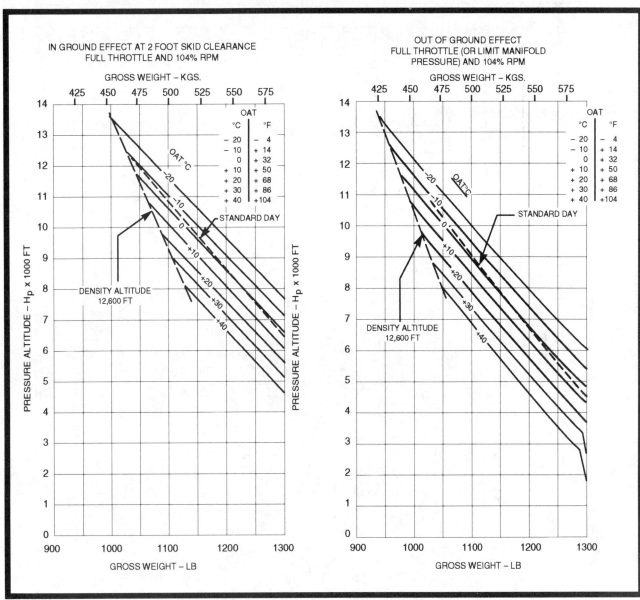

FIGURE 41.—Hover Ceiling vs. Gross Weight.

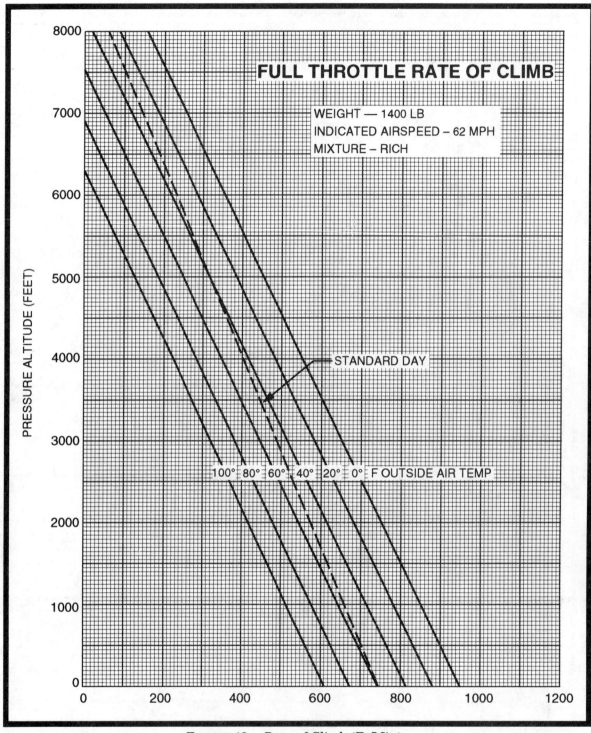

FIGURE 42.—Rate of Climb (Ft/Min).

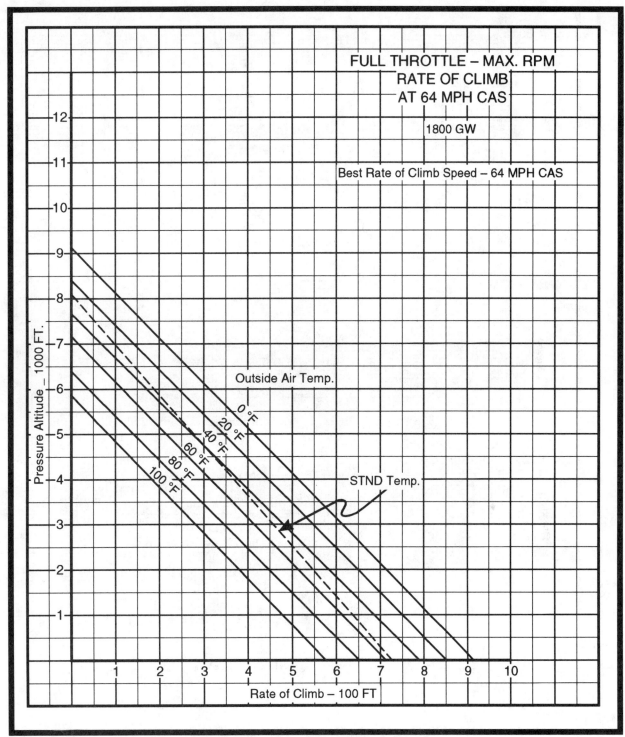

FIGURE 43.—Best Rate-of-Climb Speed.

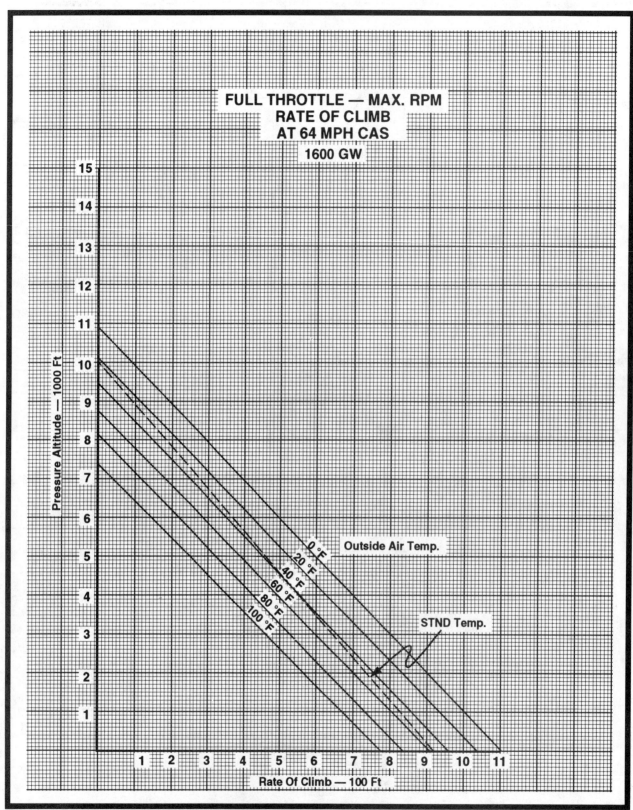

FIGURE 44.—Rate of Climb.

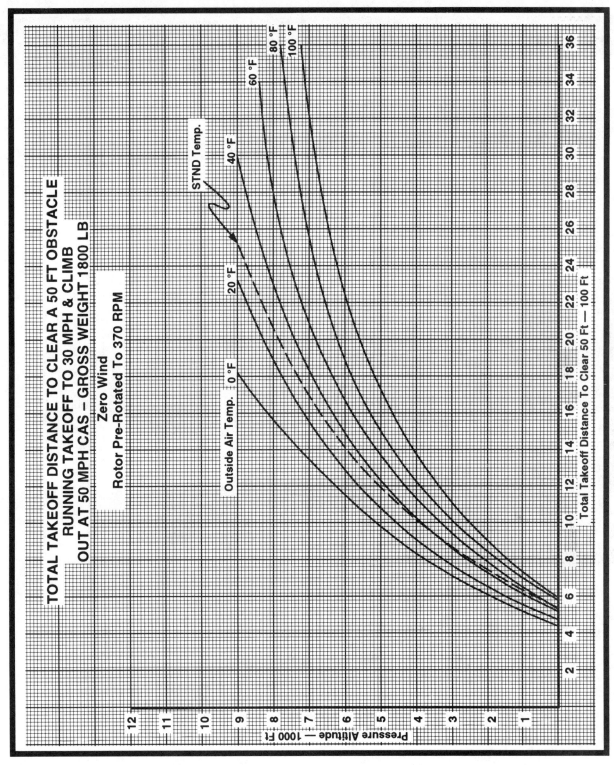

FIGURE 45.—Running Takeoff.

41

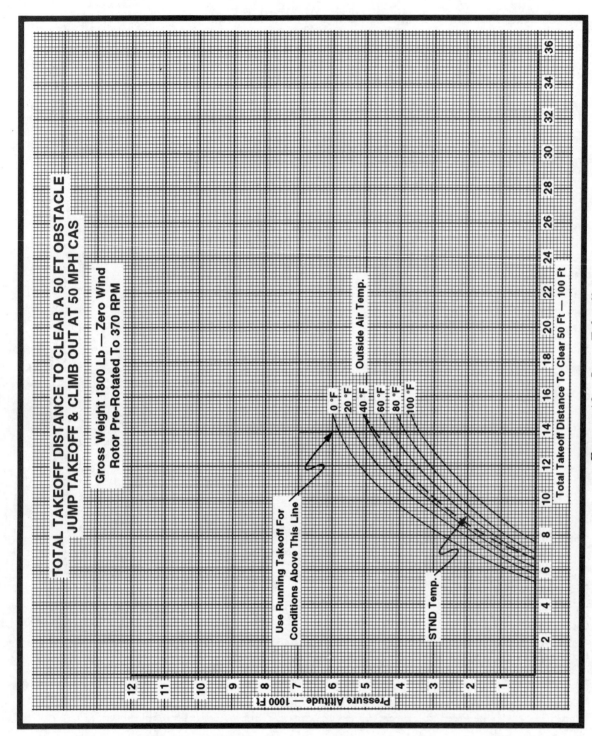

TOTAL TAKEOFF DISTANCE TO CLEAR A 50 FT OBSTACLE
JUMP TAKEOFF & CLIMB OUT AT 50 MPH CAS

Gross Weight 1800 Lb — Zero Wind
Rotor Pre-Rotated To 370 RPM

Outside Air Temp.

0 °F
20 °F
40 °F
60 °F
80 °F
100 °F

Use Running Takeoff For
Conditions Above This Line

STND Temp.

Total Takeoff Distance To Clear 50 Ft — 100 Ft

Pressure Altitude — 1000 Ft

12 11 10 9 8 7 6 5 4 3 2 1

2 4 6 8 10 12 14 16 18 20 22 24 26 28 30 32 34 36

FIGURE 46.—Jump Takeoff.

42

THIS PAGE INTENTIONALLY LEFT BLANK

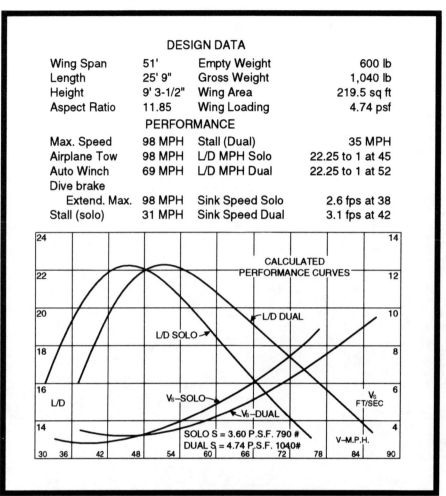

DESIGN DATA

Wing Span	51'	Empty Weight	600 lb
Length	25' 9"	Gross Weight	1,040 lb
Height	9' 3-1/2"	Wing Area	219.5 sq ft
Aspect Ratio	11.85	Wing Loading	4.74 psf

PERFORMANCE

Max. Speed	98 MPH	Stall (Dual)	35 MPH
Airplane Tow	98 MPH	L/D MPH Solo	22.25 to 1 at 45
Auto Winch	69 MPH	L/D MPH Dual	22.25 to 1 at 52
Dive brake			
Extend. Max.	98 MPH	Sink Speed Solo	2.6 fps at 38
Stall (solo)	31 MPH	Sink Speed Dual	3.1 fps at 42

CALCULATED
PERFORMANCE CURVES

L/D DUAL

L/D SOLO

L/D

Vs—SOLO

Vs—DUAL

Vs
FT/SEC

SOLO S = 3.60 P.S.F. 790 #
DUAL S = 4.74 P.S.F. 1040#

V—M.P.H.

FIGURE 48.—Performance Curves Chart.

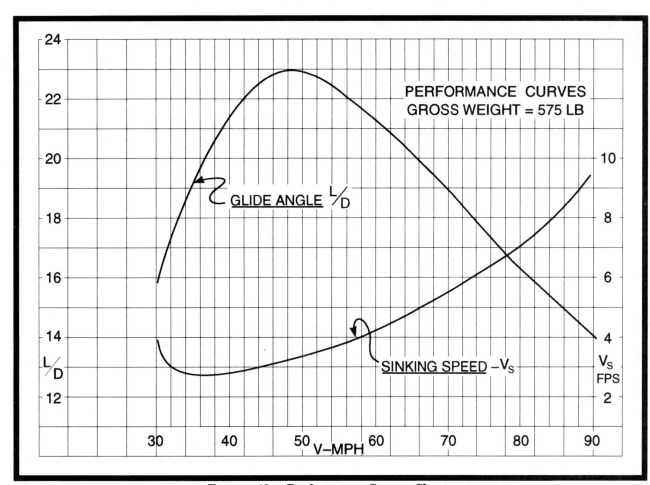

FIGURE 49.—Performance Curves Chart.

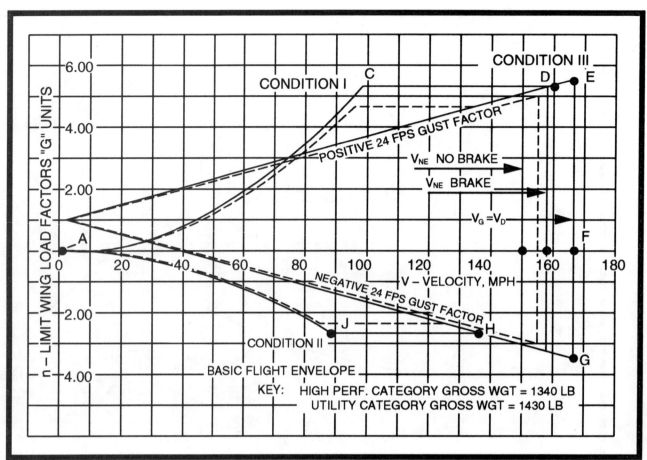

FIGURE 50.—Flight Envelope.

46

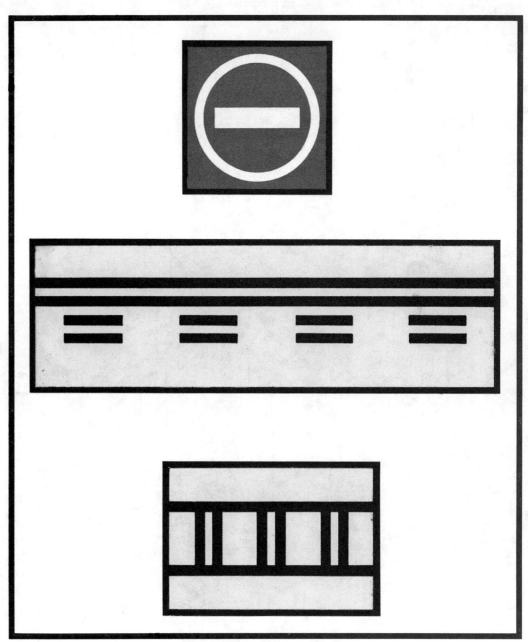

FIGURE 51.—Airport Signs.

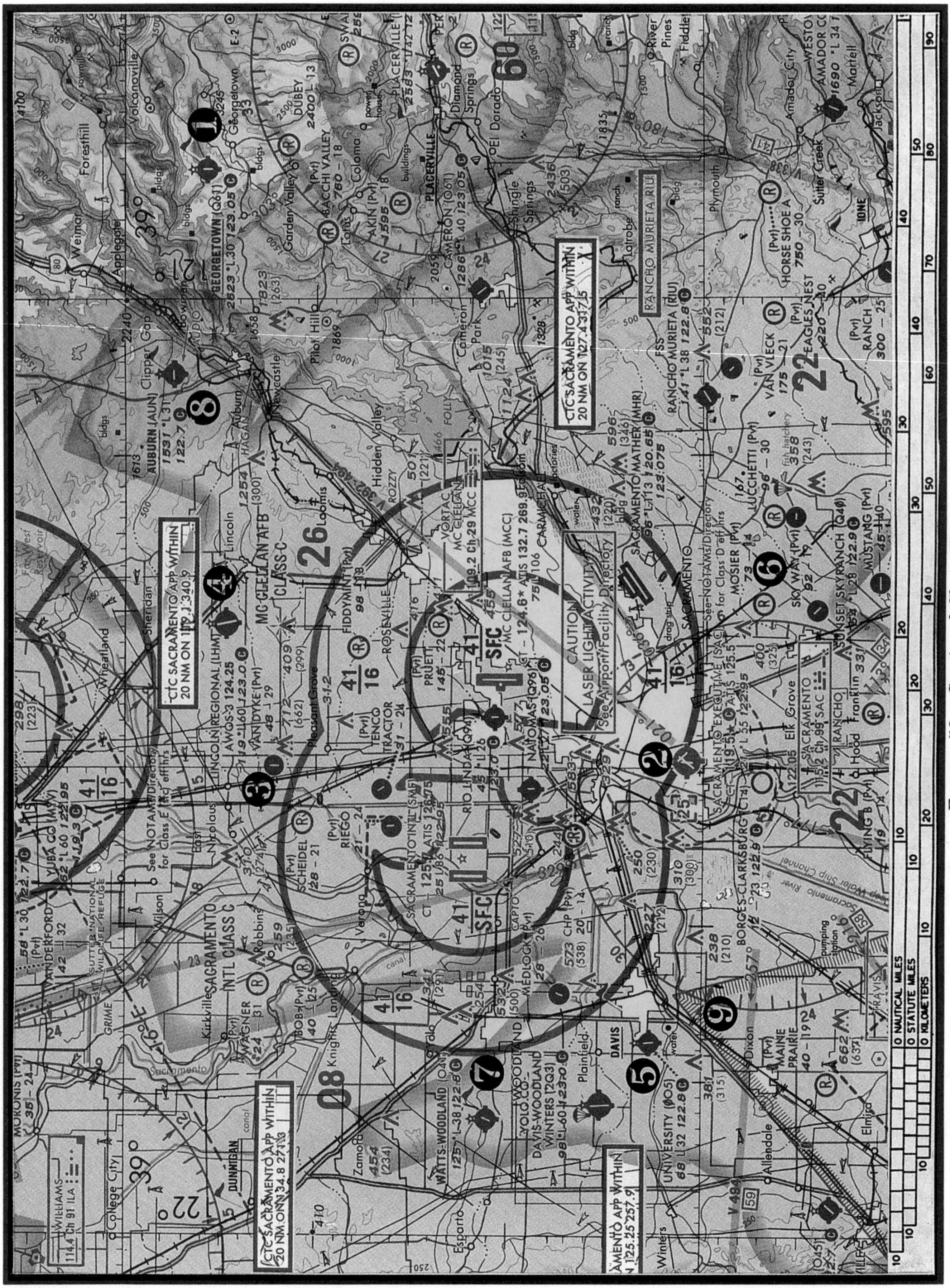

FIGURE 52.—Sectional Chart Excerpt.

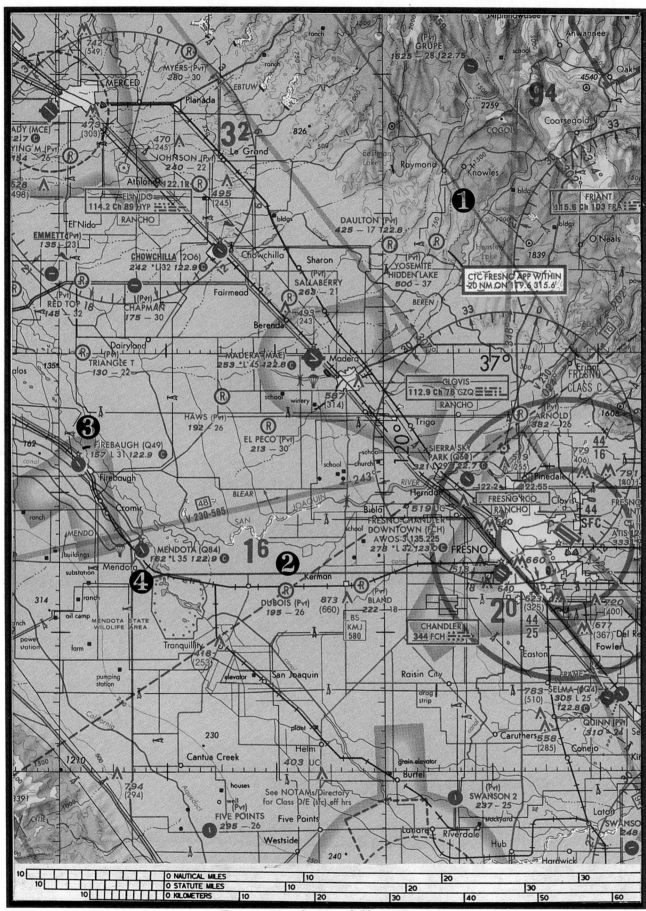

FIGURE 53.—Sectional Chart Excerpt.

49

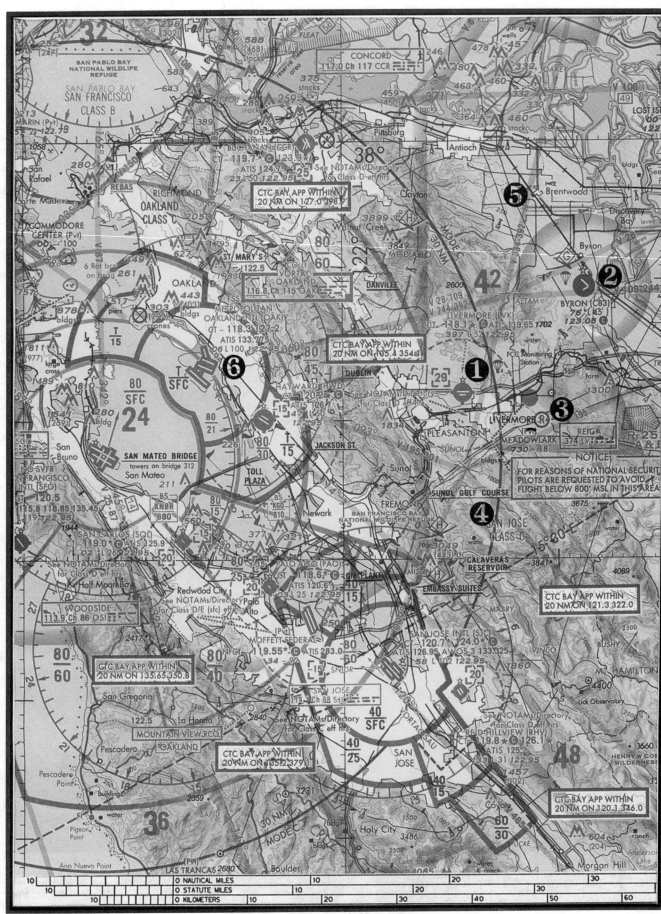

FIGURE 54.—Sectional Chart Excerpt.

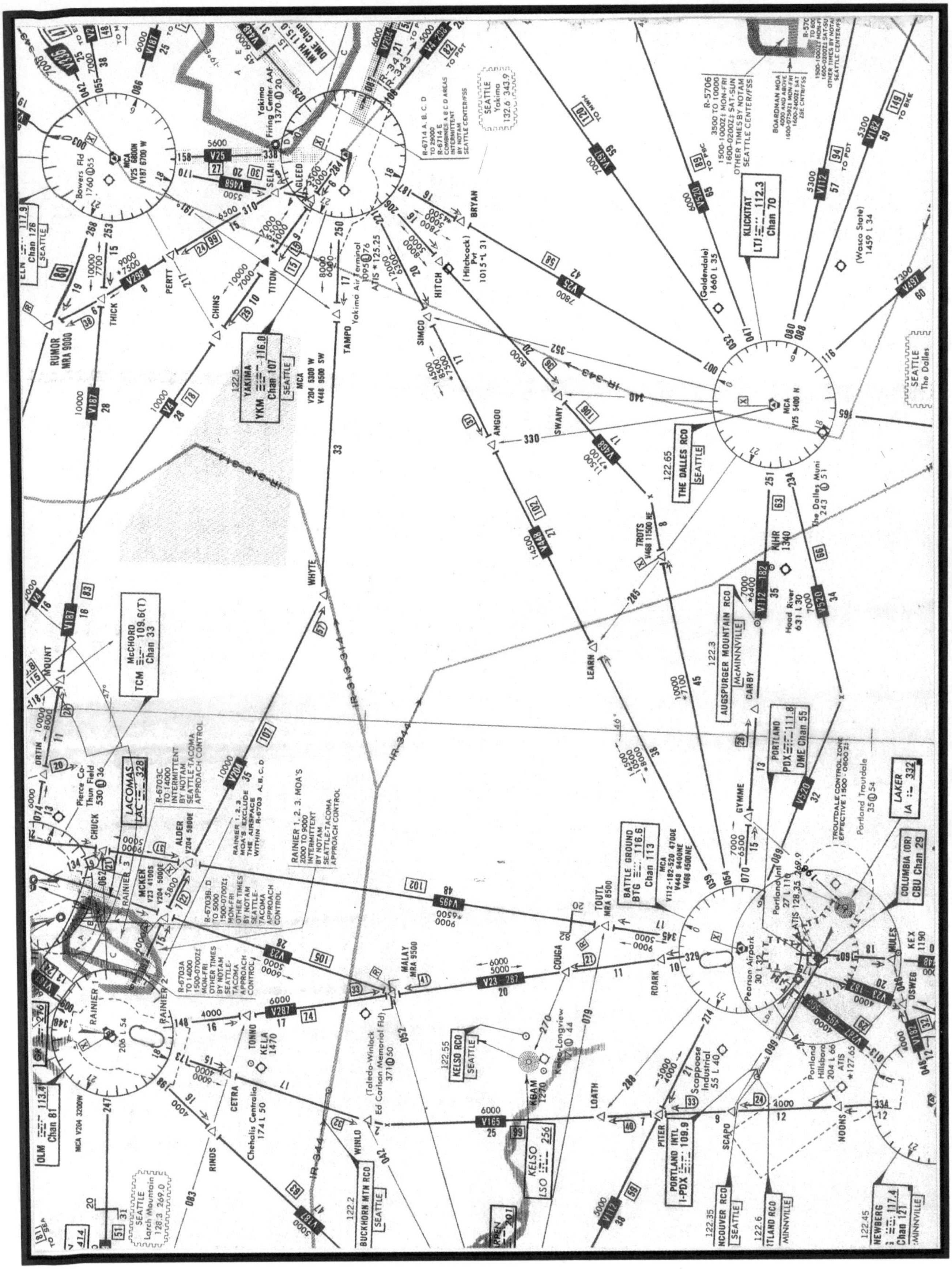

FIGURE 55.—En Route Low Altitude Chart Segment.